SPACE CHALLENGER
The Story of Guion Bluford

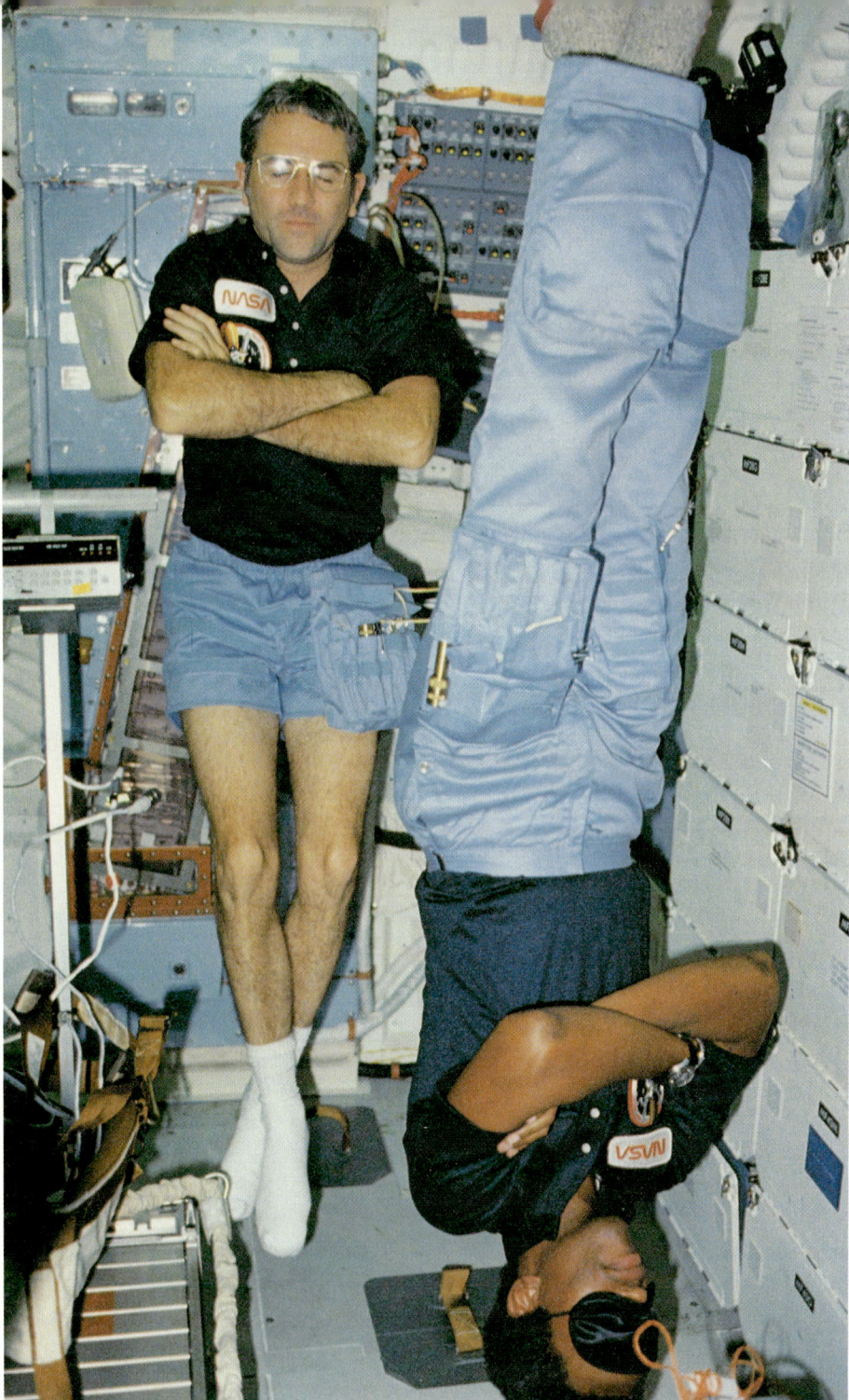

SPACE CHALLENGER

The Story of Guion Bluford

an authorized biography by
Jim Haskins and Kathleen Benson

HOUGHTON MIFFLIN COMPANY BOSTON

Atlanta   Dallas   Geneva, Illinois   Palo Alto   Princeton   Toronto

All photographs courtesy of the National Aeronautics and
Space Administration.

Space Challenger: The Story of Guion Bluford by Jim Haskins and Kathleen Benson.
Copyright ©1984. Reprinted by permission of the publisher, Carolrhoda Books, Inc., 241
First Avenue North, Minneapolis, MN 55401.

1995 Impression
Houghton Mifflin Edition, 1993

No part of this work may be reproduced or transmitted in any form or by any means,
electronic or mechanical, including photocopying and recording, or by any information
storage or retrieval system without the prior written permission of the copyright owner
unless such copying is expressly permitted by federal copyright law. With the exception
of non-profit transcription in Braille, Houghton Mifflin is not authorized to grant
permission for further uses of this work. Permission must be obtained from the
individual copyright owner as identified herein.

Printed in the U.S.A.
ISBN: 0-395-61820-7

456789-B-96 95 94

Contents

I. First Black American in Space ... 7

II. An Interest in Things That Fly ... 11

III. Not "College Material" 15

IV. Pilot and Ph.D. 19

V. Guy Bluford, Astronaut 37

VI. The Chosen One 43

VII. Aboard *Challenger* 46

VIII. A Place in History 52

    Glossary 57

    Index 62

I. First Black American in Space

Takeoff minus 10 minutes and counting. Like a glowing monument, the 100-ton space shuttle *Challenger* stood over five stories tall on launchpad 39-A. It was 1:50 A.M. on August 30, 1983. At Cape Canaveral, Florida, the thunderstorms had passed, but the air was damp and hot. Just a few miles away, alligators went about their nighttime hunting in the Banana River.

On the shore of the Atlantic Ocean where *Challenger* stood, five men finished checking their equipment and prepared for takeoff. They had now been inside *Challenger* for over an hour and had started climbing into the shuttle nearly two hours earlier. Helped by an astronaut already inside, they had climbed in and been strapped into their seats. They had then gone through a series of checks of the various shuttle systems. Then they had powered

First Black American in Space

up the auxiliary power units. Now they were ready to go. In minutes they would be the first shuttle astronauts to launch their craft at night. They would also be the first American astronauts to attempt a nighttime launch since *Apollo 17* had blasted off to the moon back in 1972.

The watching world knew that there was another "first" about this mission. One of the five astronauts was 40-year-old Lt. Colonel Guion S. Bluford, Jr., and he would soon be the first black American astronaut in space. When President Ronald Reagan sent a message to the crew before lift-off, he said, "With this effort, we acknowledge proudly the first ascent of a black American into space."

T-minus 30 seconds and counting. Lt. Colonel Bluford was not thinking about being the first black American in space. He was thinking about leaving the planet earth for the first time. He was thinking about going higher into the sky than he had ever been before and higher than most of us will ever go. But he was not afraid. "We'd spent so much time training for the mission and riding in shuttle simulators that we were pretty well prepared," says Bluford. "It's like preparing for an exam. You study as much as you can, and the better prepared you are, the less frightened you are about taking the exam."

"Lift-off!" shouted Mission Control. In a blaze of fire and smoke, *Challenger*'s twin solid-fuel rocket

First Black American in Space

boosters lifted the craft from the launchpad.
The temperature around the pad was 6000°F!
The exhaust flames were so bright that the people who had gathered to watch the launch could read newspapers at 2:30 in the morning. For the crew inside *Challenger*, it was like being in the middle of a bonfire, or riding an elevator through roaring flames.

"But otherwise, there weren't any surprises," says Bluford. "What amazed me was that the shuttle flew just like the simulator said it was going to fly. The only differences were the motion, the vibration, and the noise. You don't get those in the simulators. When I felt the movement and heard the noise, I thought, Hey, this thing really does take off and roar!"

Gradually the astronauts began to leave the incredible brightness behind. *Challenger* climbed until it was 180 miles above the earth, then went smoothly into orbit. Its crew set about doing the work they had been trained to do in this "fantastic flying machine." But all the while each was thinking, I am in space...I am orbiting the earth...I am one of the few human beings who has ever done this. And Guion Bluford was thinking of himself not as a black man, but simply as a lucky human being.

That is the way he was brought up. His parents had taught him that he was as good as anyone else. They had also taught him that if he worked hard

First Black American in Space

and tried hard he could do anything he wanted. These were lessons he had learned well. Being an astronaut was something he had begun working for and believing in many years earlier.

II. An Interest in Things That Fly

Guion Stewart Bluford, Jr., was born on November 22, 1942, in Philadelphia, Pennsylvania. His name is pronounced Guy-on, and from childhood he has preferred to be called Guy. His mother taught special education classes in the Philadelphia public school system. His father, a mechanical engineer, designed machines. Young Guy took more after his father than after his mother. He didn't mind going to school, but he was much more interested in learning on his own, and what he wanted to learn about was how things work.

The oldest of three boys, Guy had lots of mechanical toys to take apart and put back together, but what interested him most were things that fly. His brothers, Eugene and Kenneth, did not share his interest. Guy built model airplanes and collected pictures of real airplanes. When he played table tennis, he studied the way the light ball traveled

An Interest in Things That Fly

through the air and tried different ways of hitting it with the paddle to make it fly differently. He became an excellent table tennis player.

On his paper route, Guy tried different ways of folding and throwing the newspapers he delivered every day. He was a Cub Scout and then a Boy Scout, eventually working his way up to the rank of Star Scout. Although that rank has nothing to do with the stars in the sky, it seems fitting that this future astronaut was a *Star* Scout.

Many young boys are interested in model airplanes. More often than not, they want to be pilots. When Guy Bluford was growing up, there were not many black pilots. There were none at all on the few commercial airlines. In those days most people took trains and buses. They sailed to Europe on ships. Most American pilots were in the Air Force, and almost all of them were white.

That is not to say that there were no blacks who could pilot planes. The first black American to get a pilot's license was a woman. Bessie Coleman got her license in 1922, but she had to go to France for her training. The Bessie Coleman School, founded in her honor, trained other black pilots at home in the United States. By 1941, a year before Guy Bluford was born, over 100 black Americans had pilot's licenses. Japan attacked Pearl Harbor, Hawaii, in December 1941, and the United States entered World War II. During the war there was an all-black bomber squadron that saw much action in

An Interest in Things That Fly

Europe and won many medals for bravery.

So, if young Guy Bluford had wanted to be a pilot, he could at least have dreamed about it. Maybe there were not many black pilots, but there had been some. Besides, he was brought up to believe that anything was possible for him. "I never felt limited as a black person," says Guy. "My brothers and I couldn't come home and say we'd done poorly at school because we were black. My parents wouldn't have let us get away with that. They would have said, 'You did poorly because you didn't work hard enough.'"

Guy was raised in an integrated Philadelphia community and went to integrated schools. While he was growing up, he never felt different because he was black. When he was in junior high school, Little Rock, Arkansas, was ordered by a federal court to integrate its schools, and there was a furor in that community. Guy read about it in the papers and remembers, "I couldn't figure out what was going on. I remember friends of the family visiting from the South and asking me if I went to a black school. I didn't know what they were talking about, because I just went to the local school." Not until Guy reached college did he feel at all different because he was black.

Guy Bluford could have dreamed about becoming a pilot, but he didn't want to be a pilot. He didn't want to *fly* planes, he wanted to *build* them. "My room had airplane models and airplane pictures all

An Interest in Things That Fly

over the place," he says. "My interest wasn't so much in flying them, but in designing them. I was fascinated with how they were put together, why they flew. My father encouraged my inquisitiveness. He had all these engineering books at home, and I was welcome to thumb through them. I decided very early that I wanted to go into the aviation business, and I wanted to be an aerospace engineer."

III. Not "College Material"

Even now, most grown-ups don't know what an aerospace engineer does. He or she works on the design, construction, and operation of spacecraft. Back in the 1950s, when Guy Bluford was in junior high school, aerospace engineering was still a very small, and very new, field. No one had yet gone up into space, but scientists and engineers believed it was possible. According to their theories of space mechanics, spaceflight was no longer possible only in science fiction stories. Scientists in both the United States and the Soviet Union were working on artificial earth satellites. Launched into orbit around the earth, these satellites would find out more about the nature of space. They would find out if it was indeed possible for human beings to go up into space. Young Guy Bluford was excited about machines like these artificial satellites. When he grew up, he wanted to help build them.

Not "College Material"

Guy was not quite 15 years old when the Soviet Union launched *Sputnik I* in 1957. The 184-pound capsule managed to achieve earth orbit and orbited the planet for several months. That was the beginning of the "space age."

The United States did not like the idea that Russia had scored this "first." The U.S. government stepped up its space program. In 1958 the National Aeronautics and Space Administration (NASA) was created. NASA began to work on sending unmanned rockets into space. In those early days, there were a lot of mistakes. Launchings of unmanned rockets from Cape Canaveral, Florida, were often delayed. Once they were launched, they sometimes took off in the wrong direction and had to be destroyed in midair. There was a missile called the Snark. It plunged into the nearby Banana River so often that news reporters started talking about "Snark-infested waters." It was pretty clear that the American space program needed more experts in math and science if the United States was going to catch up to Russia.

The leaders of the United States took a look at American schools and decided that students should study more math and science so that America could catch up to Russia in the "space race." But no one in the schools Guy Bluford attended seemed very interested in space. Either they did not know or they did not think it important that he had been dreaming about flight even before the Russians

Not "College Material"

launched *Sputnik I*. Guy's high school guidance counselors did not urge him to go to college and study aerospace engineering. They did not urge him to go to college at all. Instead, they told him he was not college material. They said Guy should go to a technical school and learn a mechanical trade.

At Overbrook High School, Guy was not a straight-A student. "I was a weak reader," he explains, "and I tended to lean toward math and science rather than toward subjects that required a lot of reading." Perhaps the counselors thought that anyone who really wanted to be an aerospace engineer had to have top grades. Perhaps they did not think a young black man could enter the field of aerospace engineering. Luckily, Guy Bluford and his parents did not pay much attention to what the counselors said. When it came time for Guy and his classmates to decide what they were going to do after graduation, Guy applied to college. And when the editors of his senior yearbook asked him what career he wanted, Guy said, "Aerospace engineer."

"I really wasn't too concerned about what that counselor said. I just ignored it," says Guy. "I'm pretty sure that all of us have had times when somebody told us we couldn't do this or shouldn't do that. I had such a strong interest in aerospace engineering by then that nothing a counselor said was going to stop me."

There had never been any question in the Bluford household that all three boys would go to

Not "College Material"

college. "There was no way I could have gotten out of going," says Guy. Both his parents had master's degrees, and he had grandparents on both sides who'd gone to college. Guy always assumed that he *had* to go. He remembers that as he was about to leave for college his mother realized, "We never asked you if you wanted to go to college or not!"

IV. Pilot and Ph.D.

Guy was accepted into the aerospace engineering program at Pennsylvania State University. He enrolled at Penn State in the fall of 1960. "When I got to Penn State, I began to recognize that I was different because I was black," says Guy. There were only about 400 blacks in a student body of several thousand.

The civil rights movement, which had begun in the late 1950s, was in full swing by the time Guy entered college. Across the South, black people were marching and demonstrating for an end to segregation and discrimination. Many students from the North, both blacks and whites, went south to join the movement, but Guy was not among them. Although he supported the movement, he put his education first. "I was up to my elbows with studying—calculus, aerospace engineering, etcetera. I worked awfully hard at it," he explains.

Pilot and Ph.D.

In April 1961 the Soviet Union sent a man into space. That first cosmonaut, Yuri Gagarin, orbited the earth in the Russian spacecraft *Vostok I*. Twenty-three days later, the United States launched its first man into space. Commander Alan Shepard's flight in *Freedom 7* lasted just 15 minutes. He went up into space, and then he came right back down. But Americans were proud of the event. They hoped that the United States could catch up to Russia in the space race.

Most Americans were excited about the *man* who had gone up into space. Guy Bluford was excited about the *craft* that had taken him there. He wanted to learn how *Freedom 7* worked. He wanted to design a craft that would work even better. Ever since he had started building model airplanes as a child, Guy's main interest had been in flying machines. But by the time he graduated from Penn State, he had also become interested in flying the flying machines.

When Guy enrolled at Penn State, male students had to join a Reserve Officers Training Corps (ROTC) program for two years. Guy chose Air Force ROTC. At the end of his sophomore year, he could have left the program, but he chose to remain in it. "It gave me an opportunity to serve my country," he explains. "Also, the Vietnam War was starting, and I didn't want to be out on the West Coast working for some aircraft company and get drafted."

Pilot and Ph.D.

In his junior year, Guy failed a flight physical and so could not qualify as a pilot. He did not want to be a navigator, so he chose to spend his required four years in the Air Force after college as an engineer. During the summer after his junior year, he went to boot camp at Otis Air Force Base on Cape Cod, Massachusetts, and there he did pass the flight physical. He also got his first ride in an Air Force T-33 plane. "I changed directions right then and there," he says. "I decided to go into the Air Force as a pilot. I thought that if I were a pilot, I would be a better engineer."

In his senior year at Penn State, Guy was a "full bird," or pilot, in Air Force ROTC. When he graduated, he received the ROTC's Distinguished Graduate Award.

While still at Penn State, Guy married Linda Tull, a fellow student. In June 1964, shortly after Guy graduated with a bachelor of science degree, their first son, Guion III, was born. Guy immediately joined the Air Force. With his wife and son, he moved to Arizona for pilot training at Williams Air Force Base. He received his pilot wings in 1965, the same year his and Linda's second son, James T., was born.

Guy was not around much during his sons' early years. By the time he received his pilot wings, the United States was deeply involved in the war in Vietnam. North Vietnam and South Vietnam were fighting against each other. Communist China was

Pilot and Ph.D.

helping North Vietnam. The United States had stepped in to help South Vietnam. Guy Bluford joined an F-4 fighter squadron and served with the 557th Tactical Fighter Squadron based in Cam Ranh Bay, South Vietnam. Over the next few years, he flew an F-4C Phantom jet on 144 combat missions. Sixty-five of them were over enemy territory in North Vietnam. Altogether he logged some 3,000 hours of flying time and received 10 Air Force medals. His sons, says Guy, "were so young they don't even remember me being gone. But when I got back, they looked at me and said, 'Who's this guy?' There was a period when they had to adjust to the fact that they had two parents, not just one."

When Guy returned to the United States, he was considered one of the best pilots in the Air Force. In fact, the Air Force sent him to Sheppard Air Force Base in Texas to teach cross-country and acrobatic flying. But Guy had not forgotten his dream of designing aircraft. Still hoping to become an aerospace engineer, he applied to the Air Force Institute of Technology. Many other Air Force pilots wanted to attend the institute, and Guy faced stiff competition, but in 1972 he was accepted. Two years later he graduated with distinction, and with a master's degree in aerospace engineering.

He then went to work at the Air Force Flight Dynamics Laboratory at Wright-Patterson Air Force Base in Ohio. "We did a lot of testing of new airplanes and new aircraft designs," he says. "We

Pilot and Ph.D.

looked for ways to make airplanes fly faster and higher. At last I had achieved the goal I'd had as a kid—I was an aerospace engineer."

All the while, Guy was taking further courses at the Air Force Institute of Technology. In 1978 he received his Ph.D. in aerospace engineering with a minor in laser physics, the study of light and energy. To earn a doctorate, you have to write a dissertation—a long paper about some subject in your field of study. Guy Bluford's paper was titled "A Numerical Solution of Supersonic and Hypersonic Viscous Flow Fields Around Thin Planar Delta Wings."

"Delta wings are triangular wings," Guy explains. "I calculated how the air goes around the wings at speeds greater than the speed of sound—three to four times the speed of sound and faster. If you had picked a place anywhere along a wing, I could have told you what the pressure, the density, and the velocity of the air was above and below that place. I developed a computer program that could do that."

Not bad for someone who was not supposed to be college material!

Astronaut Candidate

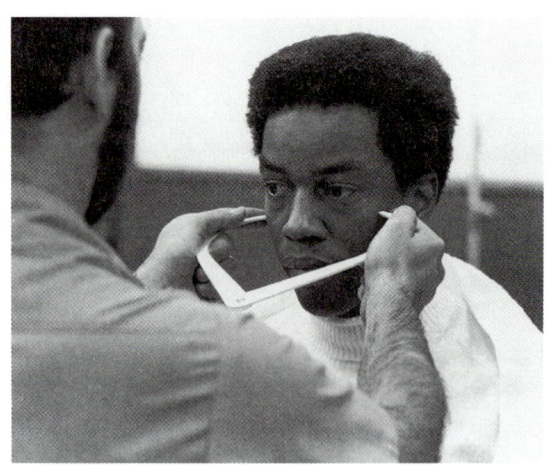

January 31, 1978. In late January and early February, Guy and the 34 other astronaut candidates visited the Johnson Space Center for orientation and such tasks as suit fitting. Here Guy is being measured for a helmet visor.

August 2, 1978. Guy prepares to begin a parasail exercise over water during water survival school at Homestead Air Force Base, Florida. A small motor boat outside the picture guides and pulls each trainee and his or her chute to simulate the feel of a parachute glide. The course was designed to prepare the trainees for possible ejection from an aircraft over water.

August 2, 1978. Guy takes part in an exercise to prepare him for retrieval by a helicopter following a parachute landing in water.

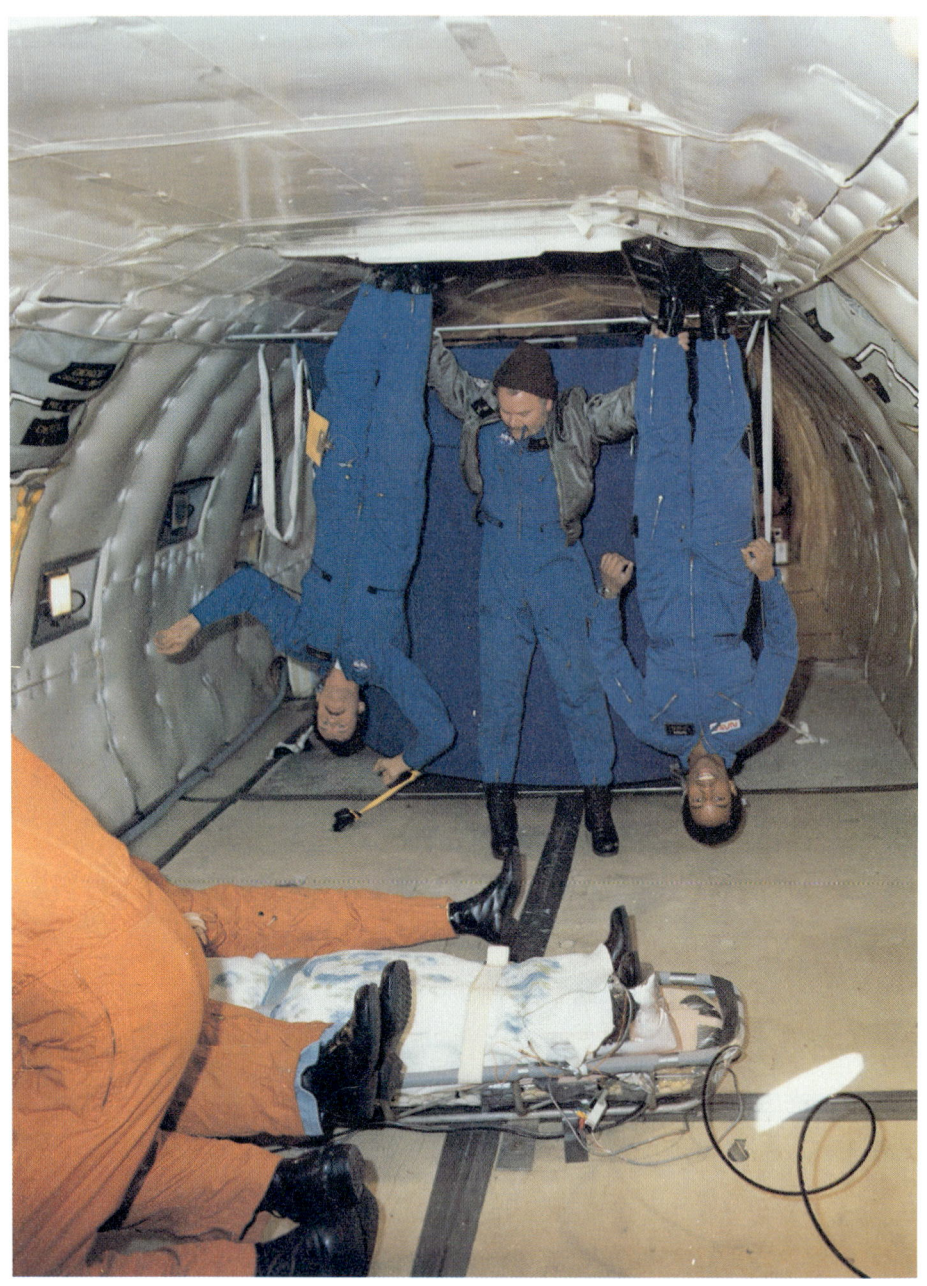

March 2, 1979. Guy gets a new view on the environment during a zero-gravity flight. He is aboard a KC-135 aircraft which repeatedly flies in a special pattern in order to provide a series of half-minute sessions of weightlessness.

The Crew

April 29, 1983. The crew members. Richard H. Truly (center) is crew commander. Daniel C. Brandenstein (left) is pilot. The mission specialists are Dale A. Gardner, William E. Thornton (both in back row), and Guion S. Bluford.

Astronauts Guion S. Bluford (right) and Daniel C. Brandenstein man their respective *Challenger* entry and ascent stations in the shuttle mission simulator at the Johnson Space Center. This photo was taken by Otis Imboden.

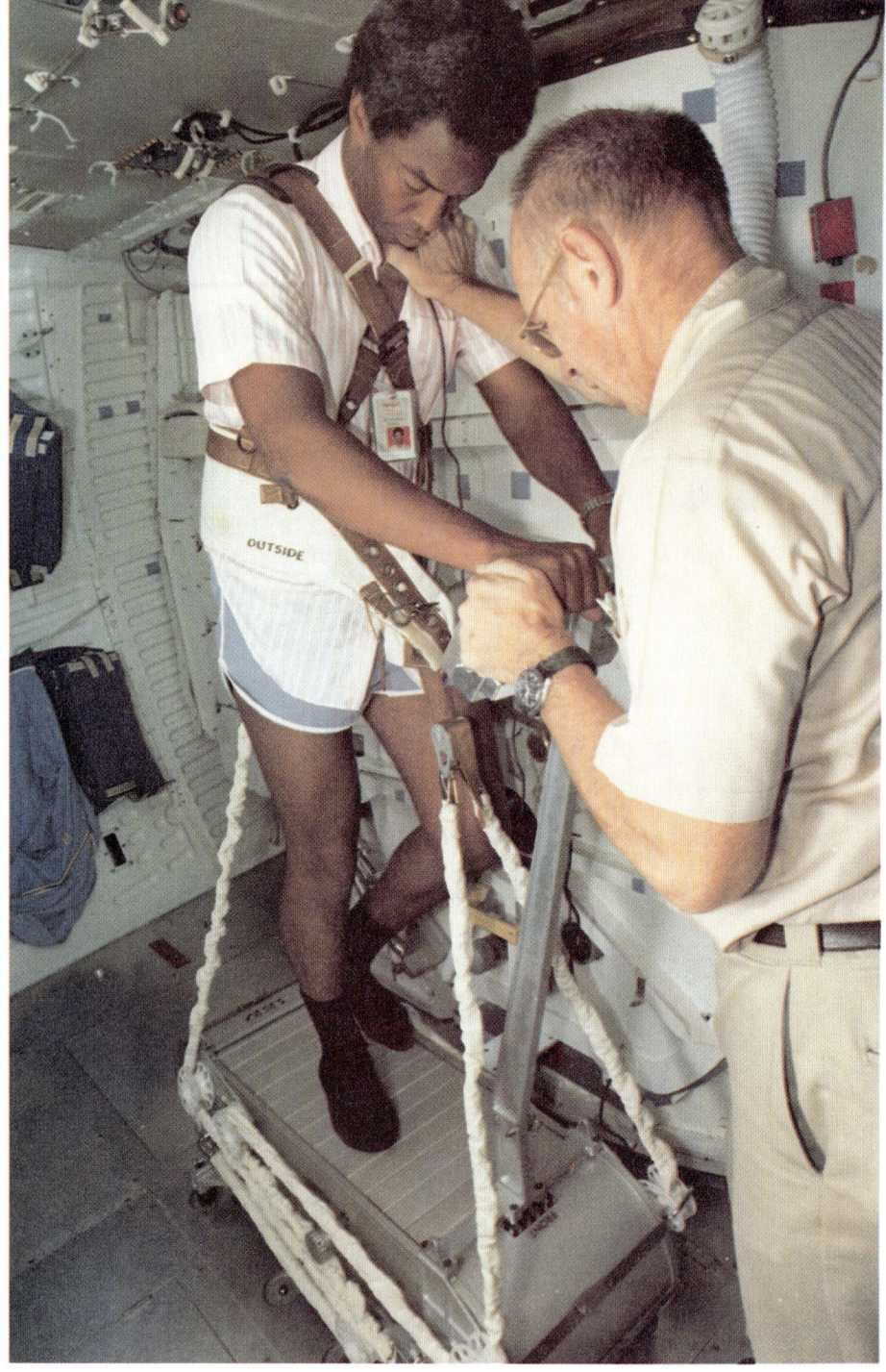

July 26, 1983. Guy and fellow mission specialist Dr. William E. Thornton demonstrate an in-orbit experiment in the Johnson Space Center's one-G trainer. Photo by Otis Imboden.

The Craft

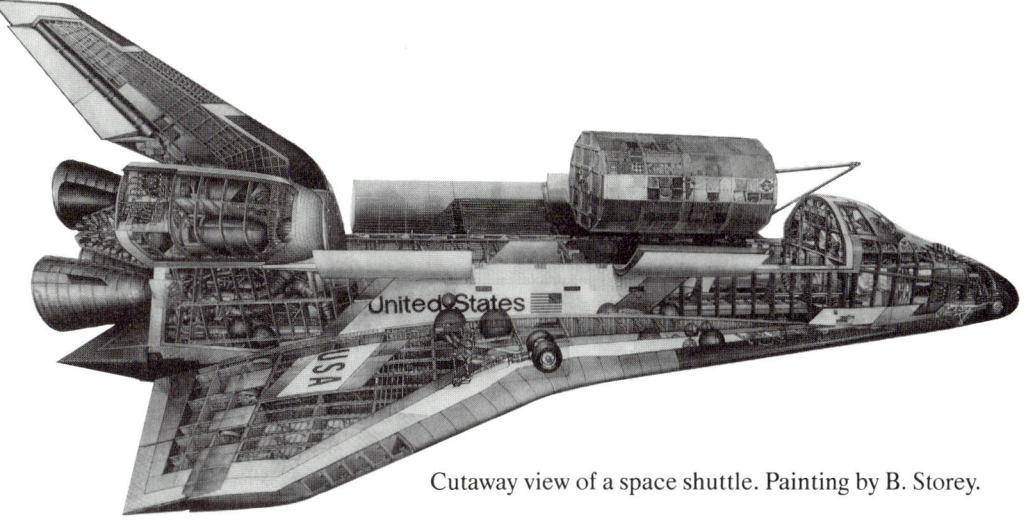

Cutaway view of a space shuttle. Painting by B. Storey.

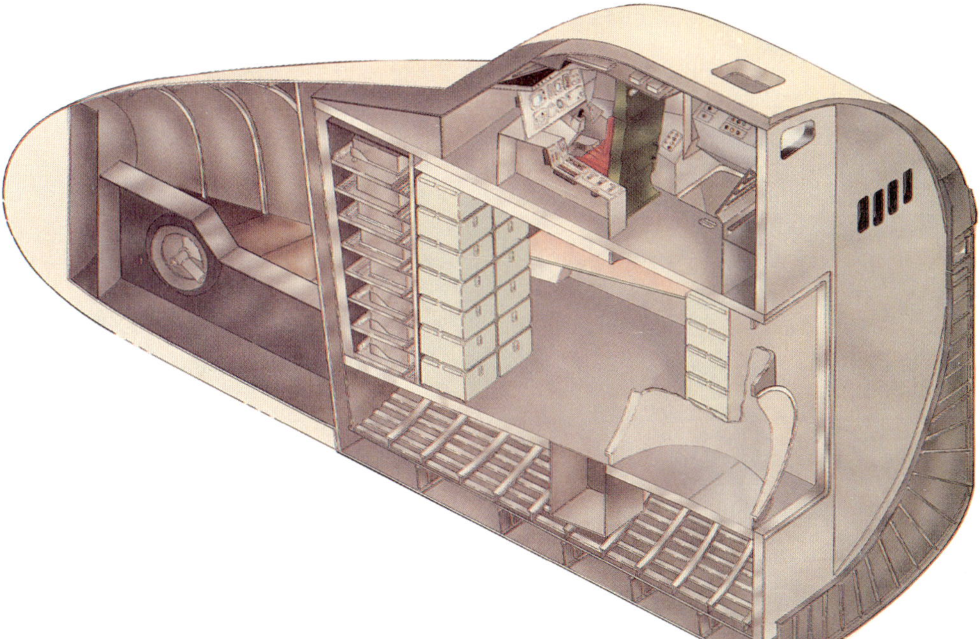

Cutaway view of the crew compartment of a space shuttle. This view is of the starboard (right) side of the module, looking forward.

An interior view of the upper flight deck of the crew compartment of a space shuttle. The crew is seated in their launch positions.

A cutaway view of a space shuttle's crew compartment. The crewmen are at their in-orbit work stations in the upper flight deck. Note the open doors of the huge cargo bay, at right.

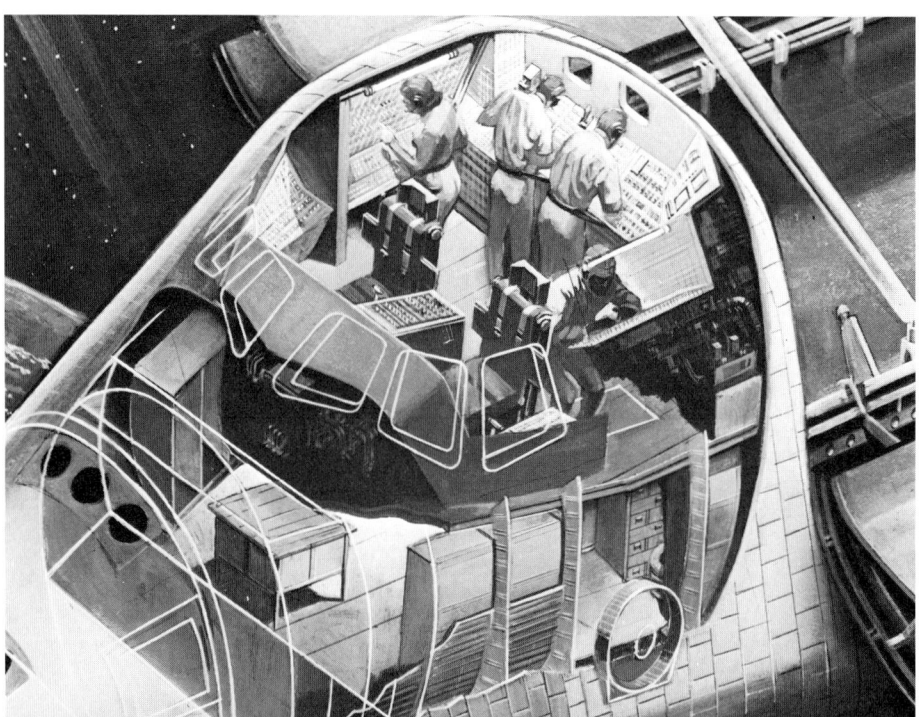

Astronauts eating a meal in the shuttle's middeck

Four crew members go through a training exercise at the Johnson Space Center. Their seating pattern shows launch and landing positions.

The Flight

Guy checks one of the control knobs on the continuous flow electrophoresis system experiment aboard *Challenger*.

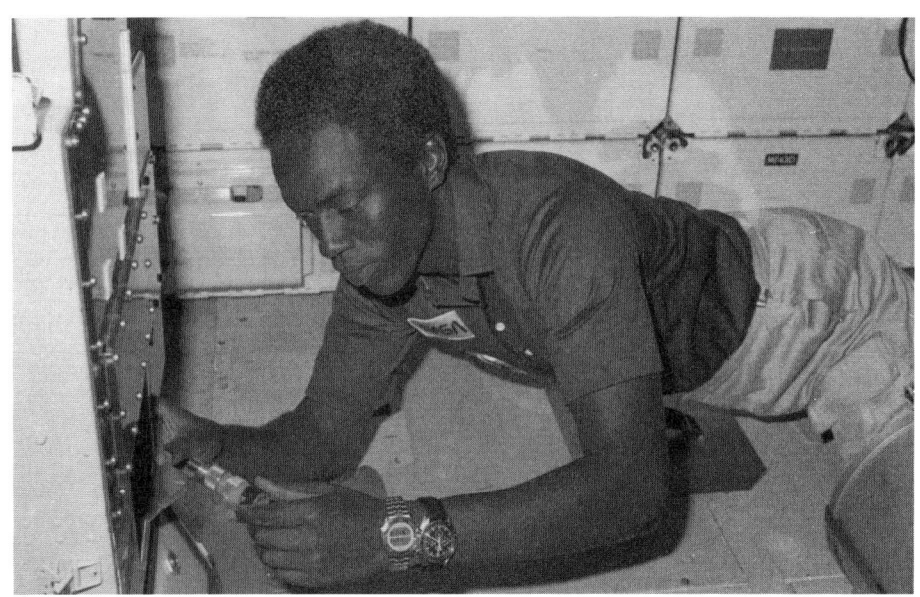

Guy at work in flight.

Dale Gardner asleep.

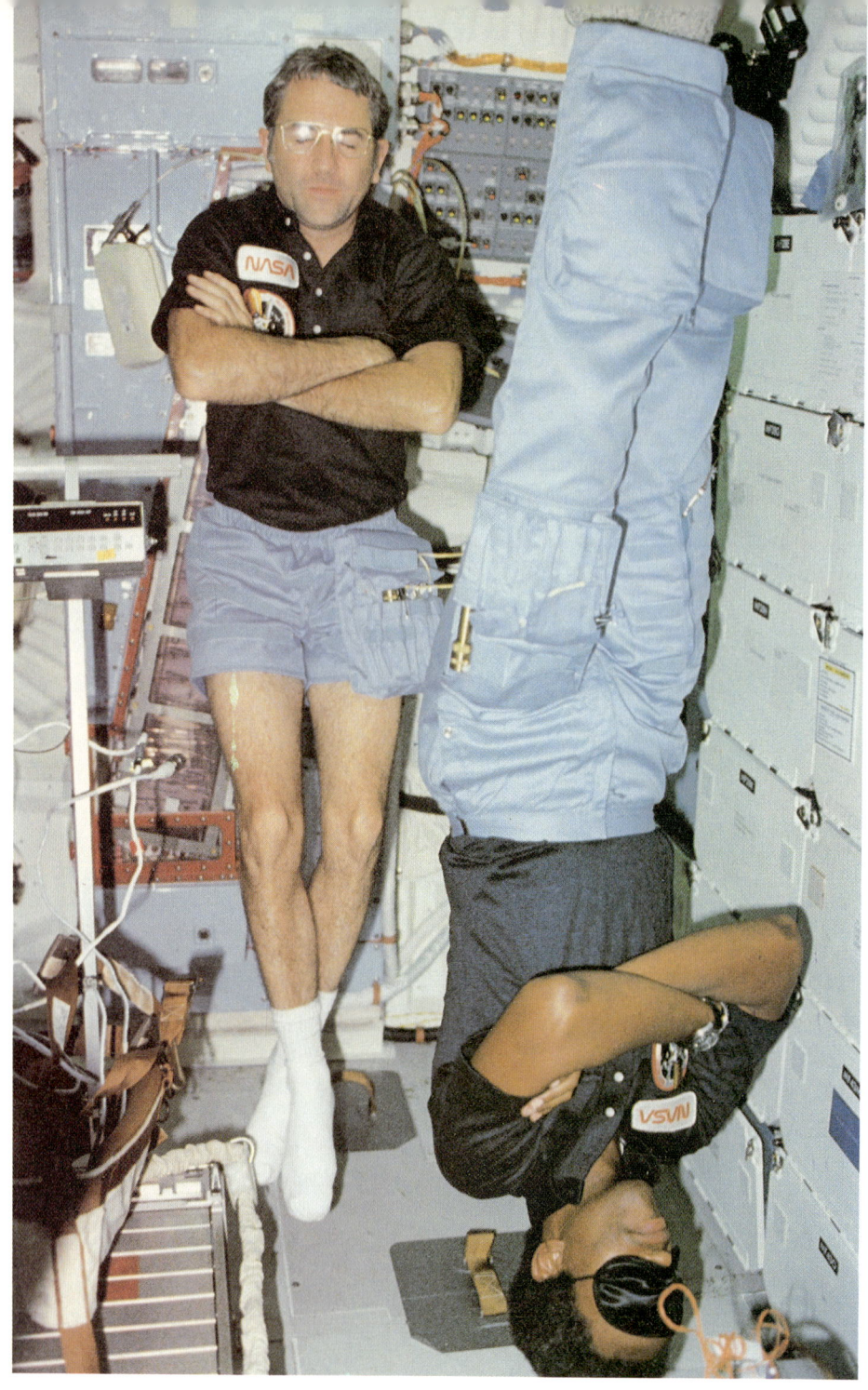

Guy and Commander Truly asleep.

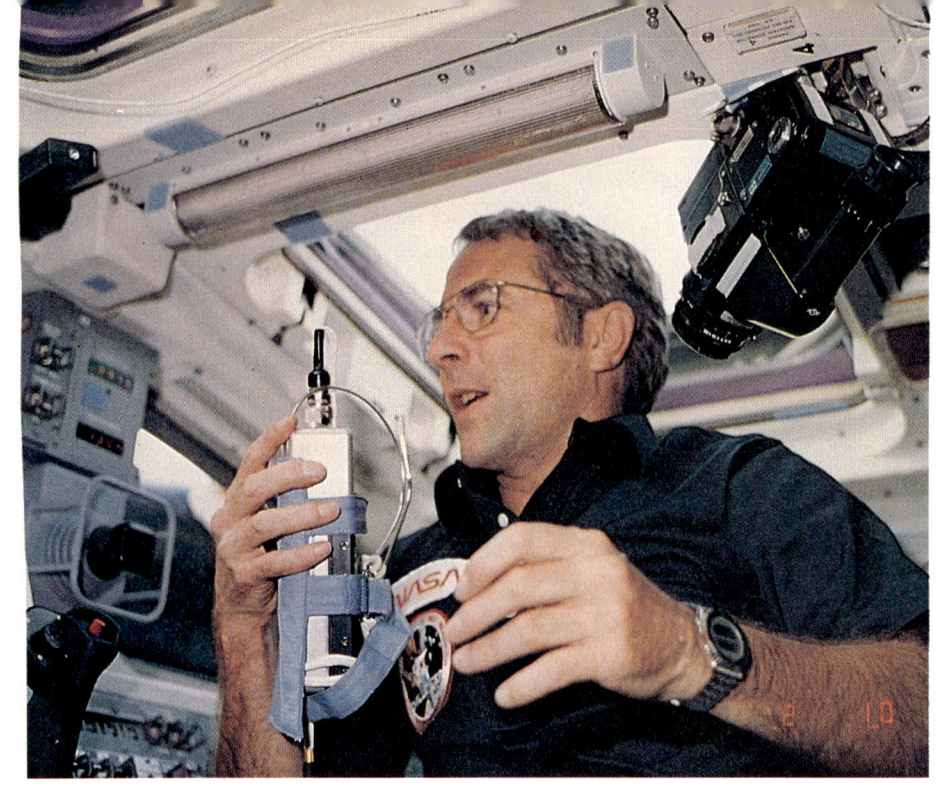

Richard H. Truly, crew commander, communicates with ground controllers from the flight deck of *Challenger*.

William Thornton (right) conducts a test on Dale Gardner in the middeck of the *Challenger*.

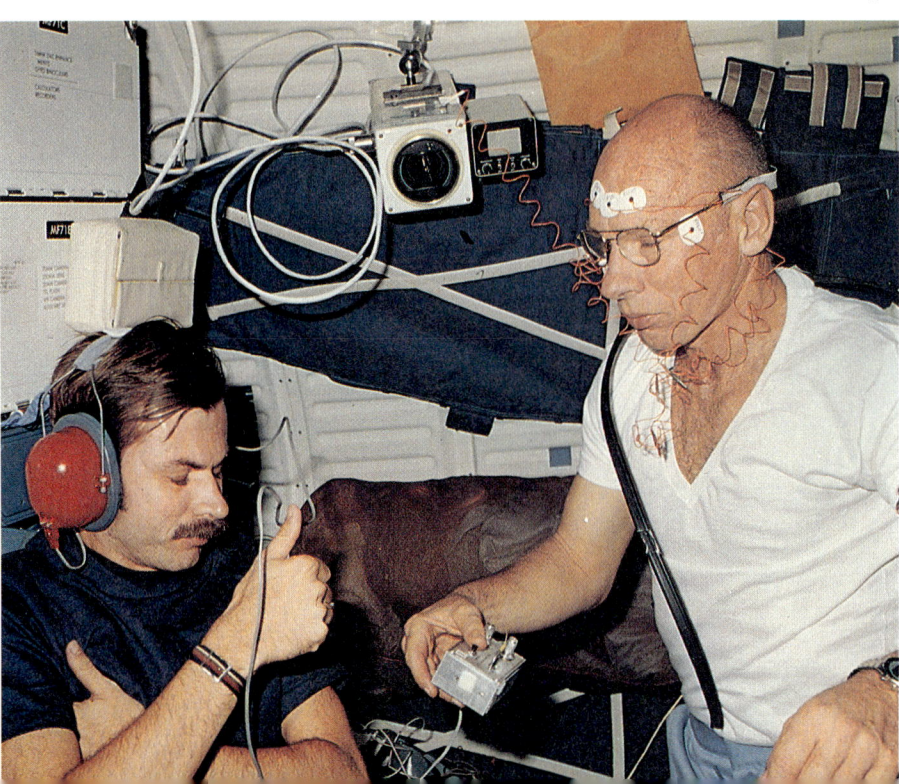

The Indian National Satellite (Insat-1B) is about to clear the earth-orbiting *Challenger*.

V. Guy Bluford, Astronaut

In 1978, the same year he received his doctorate, 35-year-old Guy Bluford applied for the astronaut program. It seemed to him that NASA was the best place to learn new things about aerospace engineering. Even though the American government and the American people did not seem as interested in space exploration as they once had been, NASA was still doing exciting things in the field. To Guy Bluford, NASA seemed the ideal place to combine his flying and engineering interests, but he did not really expect to get into the astronaut program. That year alone, 8,878 other people also applied.

One day a few weeks after he had applied, someone from NASA called him. The man started talking about the weather in Texas, and Guy was a little puzzled. Then the man asked Guy if he would like to move to Houston and work at the Johnson Space Flight Center. Guy couldn't believe it—he

Guy Bluford, Astronaut

was being invited to become an astronaut! He had been accepted into the program over thousands and thousands of other applicants. Just 35 men and women joined the astronaut program that year. One of the others was Sally Ride.

Guy Bluford moved his family to Houston, Texas, near the Johnson Space Flight Center. His wife, Linda, got a job as an accountant with an oil company. Their sons, Guion III and James T., enrolled in local high schools. Guy looked forward to working in the space program with other scientists.

Guy was not the first black American to become an astronaut. Back in the early 1960s, President John F. Kennedy had appointed the first black to the astronaut program. Captain Edward Dwight joined the program in 1962, the year after Commander Alan Shepard's 15-minute ride into space on *Freedom 7*. But Dwight left the space program in 1966. Now a retired Air Force captain, he says he quit under pressure. "They didn't want black involvement," he says. "They felt that to send blacks into space would lessen the general public's enthusiasm for the space program." The Air Force claims that Dwight failed to complete his training.

Vice President Lyndon Johnson became president after President Kennedy was assassinated. He appointed another black, Bob Lawrence, to the astronaut program. But Lawrence died in a plane crash soon afterward.

Guy Bluford, Astronaut

When Guy Bluford was accepted into the astronaut program, two other blacks were also accepted.

Guy says he did not feel any pressure because he was black. He followed the same steps in the program that other new members did. As an astronaut candidate, he was a student for a year. He attended classes for about six months, studying subjects such as shuttle systems, geology, medicine, aerodynamics, communications, and astronomy. Then he and the other astronaut candidates went on field trips. "We went to a lot of the NASA space centers," he says, "including Kennedy Space Center at Cape Canaveral, Florida, Marshall Space Flight Center in Huntsville, Alabama, where they develop the engines, and Rockwell Aircraft Company on the West Coast where they build the shuttles. We traveled around the country, meeting all of the people associated with the shuttle program."

Back in Houston, Guy spent three months working on the mechanical arm that is used to move payloads in and out of a spacecraft's cargo bay and on some of the experiments connected with the space lab program.

When the year was up, Guy and his fellow astronaut candidates became full-fledged astronauts. "Every Monday morning we would all meet in the Astronaut Office," says Guy, "and one morning while we were sitting there, John Young, who was head of the Astronaut Office, came in and said, 'Hey, you guys have been with us for a year, and we've decided to make you astronauts.' That meant we were all now eligible to fly in the space shuttle."

Guy spent the next few years flying in "shuttle simulators" (machines that imitate the look and feel of a space shuttle) both in Houston at the Johnson Space Flight Center and in California at Rockwell Aircraft Company where the shuttles were being built. "We practiced verifying the computer instructions and learned how to make the shuttle fly properly during ascents, entries, and orbital activities," says Guy.

Guy enjoyed working at the Johnson Space Flight Center. "It has really proved to be better than I had expected," he said after entering the program. "It gives me a chance to use all of my skills doing something that is pretty exciting. The job is so fantastic, I don't need a hobby. My hobby is going to work."

By the time Guy Bluford entered the astronaut program, America's interests in space had changed. During the 1960s and in the early 1970s, the space program had developed by leaps and bounds. On July 20, 1969, Neil Armstrong and Edwin Aldrin, traveling in an Apollo spacecraft, were the first men to land on the moon. In the early 1970s, the United States continued manned space exploration in *Skylab*, an earth-orbiting space station. The main capsule was launched by an unmanned booster rocket; the crews arrived later in an Apollo-type craft that docked to the main capsule.

After 1972, however, most space exploration was done by unmanned craft. The Mariner program studied Venus and Mars. The Viking program concentrated on Mars, its primary mission to search for life. The Pioneer program studied the outer planets and more distant space. In 1973, *Pioneer 10* passed through the asteroid belt to become the first man-made object to escape from the solar system. In 1974, *Pioneer 11* photographed Jupiter's moons.

Sending rockets into space cost a lot of money, especially when the rockets could only make one flight. By the middle of the 1970s, the United States government—and the people—did not want to keep spending so much money just to explore outer space. Unless the NASA scientists could develop reusable spaceships and come up with practical uses for them, the space program could not continue as it had before.

In 1972, NASA started to develop space shuttles. Shuttles are part rocket, part spacecraft, and part airplane. They are part rocket because they are equipped with rockets that propel them. They are part spacecraft because they can navigate in outer space, something many rockets cannot do—rockets that are used as weapons, for example. They are part airplane because they have delta wings and can glide on air currents. Shuttles are reusable spaceships.

Space shuttles can do work in space, such as launching communications satellites that improve communications by telephone, radio, and TV on earth. They can build space stations, which are orbiting craft where astronauts and engineers can live, operate laboratories, and launch other craft to more distant reaches of space. Shuttles can serve as "commuter lines" between space stations and earth.

The first space shuttle, *Columbia*, was launched in April 1981. On July 4, 1982, it completed its fourth flight. Then *Challenger*, America's second space shuttle, went into operation. By May 1983, it had made one flight. On its second flight in June 1983, just two months before the flight Guy Bluford would be on, one of its three mission specialists was Dr. Sally K. Ride, the first American woman in space. (The Russians had sent a woman cosmonaut into space in 1963, and another in 1982.) *Challenger*'s third flight, and the seventh mission for the shuttle program, would carry the first black American into space: Lieutenant Colonel Guion Stewart Bluford, Jr.

Guy didn't know what the top brass at NASA wanted when he was called in to see them. He thought he must have done something wrong. When they told him he had been selected for *Challenger*'s third mission, he couldn't believe it!

VI. The Chosen One

Some people pointed out that Russia had beaten the United States to this "first" too. The Soviet Union had sent a black Cuban into space in 1980. Some people wanted to know why the first American woman had been sent up into space before the first American black. But Guy Bluford did not worry about such things. "I am not slighted by Sally Ride's going first," he said. "We're all going to get an opportunity to fly. It's not when I fly but how well I fly that counts." Actually, he was relieved that Sally Ride did go first. All the attention she got would, he hoped, take some of the attention away from him.

Guy Bluford wanted to be known as a man who did a good job, not for the color of his skin. He did not want to be singled out as a "first." He often reminded people that there were 2 other black astronauts in the 78-member astronaut program.

The Chosen One

One, Ronald McNair, was scheduled to be on the shuttle mission set for early 1984. They did not resent the fact that Bluford would be first. "The three of us never talk about my being first," he said. "We all recognize that somebody has to play this role, just as one of the women had to be first. I'm just looking forward to flying."

Guy would be one of the three mission specialists on *Challenger*'s third flight. He could have been a pilot. "I'm a strange bird because I'm equally as qualified to be a pilot as I am to be a mission specialist," he says. "I just happened to go into the program as a mission specialist, and that's where I fit right now."

The shuttle is so complicated that both pilots and mission specialists are needed to operate it. The pilots fly the shuttle. The mission specialists are experts on the cargo that the shuttle is carrying. They conduct scientific experiments, launch satellites, and do all the other things the shuttle is up there to do.

Guy was to be in charge of deploying an Indian satellite. The country of India had paid for *Challenger* to launch a communications and weather satellite. Indian scientists were hoping to use it to help improve their nation's telephone system, to produce more reliable weather forecasts, and to beam educational television to rural villages.

Two American companies, McDonnell Douglas and Johnson & Johnson Pharmaceutical Company, were paying for experiments having to do with the separation of biological material with electromagnetic waves. Guy would be one of the crew members working on these experiments.

The sending of the first black American into space was an historic event. Many famous blacks went to Cape Canaveral, Florida, to watch the launching: Bill Cosby, the comedian; Wilt Chamberlain, the former basketball star; Lionel Hampton, the musician; and Dr. James Cheek, President of Howard University in Washington, D.C. (Sadly, neither of Guy's parents had lived to see their son become the first black American in space. His father died in 1967 and his mother in 1978.) When *Challenger* lifted off in a blaze of flame, however, no one was thinking about the races of the five crewmen aboard. Bill Cosby did not feel that Guy Bluford should be watched closely because he was black. "Our race is one which has been quite qualified for a long time," Cosby said. "The people who have allowed Guy to make this mission are the ones who have passed the test."

VII. Aboard Challenger

Once in orbit, Guy Bluford and his fellow crew members unstrapped themselves from their seats and began to adjust to life in "Zero G" (zero gravity). Part of their training for the mission had been to spend time in a "water immersion facility," a big tank of water that simulates something like weightlessness. But the real thing took some adjusting to.

"There is no feeling of right side up or upside down up there," says Guy. "We knew we were upside down because, even though earth was below us, we were looking at it out of our upper windows. But when you're floating around in space, you feel the same when you're upside down as you do when you're right side up. You don't feel any different when you're standing on the ceiling than you do when you're standing on the floor. Imagine being able to walk across the ceiling or along the walls

Aboard Challenger

as easily as you can walk on the floor!"

Once they had gotten somewhat used to Zero G, the mission specialists went to work. They tested a new communications link between the shuttle and Mission Control. Back in April, the first *Challenger* mission had put this tracking and data relay satellite into space.

That early Tuesday morning, they conducted two experiments with a six-foot-tall biological processing machine. They put living cells from pancreas, kidney, and pituitary glands through an electrical field in order to separate the hormone-producing cells from the others. Scientists have found that each different type of cell in both human and animal bodies has a particular electrical charge, almost like a fingerprint. By passing cells through an electrical field, they can separate out, or isolate, particular kinds of cells. This can be done on earth, but it can be done much more quickly in space, where the effect of gravity does not interfere with the separation process. The astronauts were trying to isolate insulin-producing cells from a dog pancreas. Medicines made from these cells might be used to treat diabetes. Special cells in the human kidney produce an anti-clotting substance that might be used to treat certain blood diseases. Special cells from the pituitary gland of a rat might be used to treat dwarfism. Medical scientists hope that one day the space shuttles can become laboratories where astronauts can harvest great quantities of these special cells.

Aboard Challenger

The mission specialists also checked the satellite they were to put into orbit for India.

At 1:30 P.M. they quit for the day and ate their first supper in space. They were luckier than the first astronauts, who had to eat food from tubes. They ate hot meals with knives and forks and spoons. "We had little trays with magnets to hold the metal utensils," says Guy. "The food was stored in plastic containers. It was dehydrated, so you couldn't tell if it was macaroni and cheese or roast beef unless you read the label. But once you added water to the container and heated it up, you could cut off the top of the container and have regular food, just like on earth. There was a wide variety of dishes—fruit cocktail, sandwiches, roast beef. The only restriction was that the food had to stick to your plate, like macaroni and cheese. Peas would just float away. So if you did have peas, they had to be in a sauce."

After supper, the astronauts did some space sightseeing. They could not see the whole earth because they were not up high enough. The earth is 25,000 miles around at the equator. From the *Challenger*'s position 170-180 miles away, the astronauts could see the earth's horizon as curved, but they could not see the earth as a ball. "It was like looking out the window of an airplane, only much higher up," says Guy. "Also, when you look out the window of an airplane, what you see on the ground goes by slowly. We were moving at 300 miles a

48

Aboard Challenger

minute, so for us the ground went by very quickly."

The *Challenger* circled the earth every 90 minutes. "We flew over the Florida peninsula and could see it all, plus Cuba and some of the islands in the Caribbean Sea," says Guy. But because the earth rotates, 90 minutes later the astronauts wouldn't have the same view. They would be looking down at the earth several degrees to the west, or over New Orleans.

During daylight hours in the United States, the astronauts slept. "Some of the guys went upstairs and strapped themselves into their seats," says Guy. "I didn't. I went downstairs and tied one end of a string to my waist and the other end to the lockers. Then I floated out into the middle of the room and fell asleep. If I hadn't used a tether, I would have floated all over the cockpit and bumped into things. Occasionally I'd float up against the lockers and be jarred awake. It felt funny waking up and not knowing if I was upside down or not."

Orbiting the earth every 90 minutes, the astronauts saw 45 minutes of daylight, then 45 minutes of darkness. When they slept, they wore eyeshades so that the daylight would not awaken them.

Very early Wednesday morning, the mission specialists launched the Insat-1B satellite for India. The special launching of this satellite was the reason that they had taken off from Cape Canaveral at night. In order to put the satellite in the correct position, it had to be ejected above the Pacific

Aboard Challenger

Ocean just as the sun was setting, and the *Challenger* had to be placed in an orbit different from the orbits of the six earlier shuttle flights. It was an orbit that could only be achieved through a night launch.

While orbiting 185 miles above the Pacific Ocean, the crew set the Insat-1B spinning outside the open doors of the shuttle's cargo bay. The satellite spun nearby in space for about 45 minutes. Then, propelled by its own rocket boosters, it began a smooth climb to an altitude of 22,300 miles above the earth, where it went into orbit. Guy, the mission specialist most responsible for the satellite launching, was pleased. He reported to Mission Control, "The deployment was on time, and the satellite looks good."

As they worked, Guy and the others floated about weightlessly inside the cabin. This weightlessness causes some astronauts to have motion sickness. Guy and his fellow crew members acted as guinea pigs in an attempt to find out the reason for the motion sickness. As they floated about, they wore electrodes on their foreheads and temples. These electrodes recorded their eye movements.

The mission was very routine except for one incident. The crew was startled to hear the whine of a smoke detector! A fire inside *Challenger* would have been a disaster. The crew looked to the seven fire extinguishers in the cabin. Luckily, this was a false alarm. It had been set off by a very delicate sensor in the cockpit. There was no smoke inside *Challenger*.

President Ronald Reagan placed a special telephone call to the orbiting shuttle. He praised all five members of the crew, but he had a special message for Guy Bluford. "You, I think, are paving the way for others," said the president, "and you are making it plain that we are in an era of brotherhood here in our land."

VIII. A Place in History

Their work completed, the crew of the *Challenger* rested and prepared for their landing in the early hours of September 5.

Soon after midnight, as the shuttle passed over the Indian Ocean on its 98th orbit, the two pilots, Captain Richard Truly and Lieutenant Commander Dale Gardner, fired the shuttle's two orbital maneuvering engines to slow the craft down. The shuttle descended into a lower orbit, then made its fiery plunge into the earth's atmosphere over the Pacific Ocean. Ships between Guam and Hawaii could probably have seen the spaceship, glowing from the heat of atmospheric friction, streaking across the sky. Then *Challenger* glided in over the California coast.

The paved runway at Edwards Air Force Base in California was outlined with lights. Out in the Mojave Desert, 7,500 feet from the end of the

A Place in History

runway, a system of lights showed the shuttle pilots where the correct glide path, or approach, was. When the shuttle was only 10 minutes away, no one on the ground could see it. Without lights and no longer glowing with heat, it was invisible in the sky. A few minutes later, though, the people waiting on the ground heard loud sonic booms. The shuttle was just 3 minutes away from touchdown.

Most people did not see *Challenger* until a few seconds before touchdown. Its wheels made contact with the concrete runway and it coasted to a stop. Minutes later the hatch was opened, and the five astronauts climbed out of the shuttle to cheers from the crowd.

After showering and shaving and changing their clothes, the astronauts held a brief news conference, then returned to their home base in Houston. *Challenger* stayed in California until it could be carried piggyback by a huge 747 jumbo jet back to Houston. On arrival in Houston, each astronaut was introduced to the waiting crowd of relatives and NASA workers. The loudest cheers were for Guy Bluford. He smiled and waved. "I have a deep feeling of thanks to you, because without you we wouldn't have had as much fun as we were having," he said. Then he hugged his wife and his Aunt Lucille.

For a while, the five *Challenger* astronauts had little chance to rest. In addition to the official debriefings and medical examinations, there were

A Place in History

press conferences and interviews and an appearance on a Bob Hope television special. But after a few weeks, life for the five men returned to normal. Guy Bluford was glad. He had enjoyed going up into space. He was pleased to be a role model, someone for young blacks to look to and say, "Gee, if this guy can do it, maybe I can do it too." But he is a quiet man who prefers not to be the center of attention.

"We've worked very hard not to have any hoopla," says Guy. "My family accepts what has happened and is proud of it, but no one carries it around. My kids are in college, and if you stop them on campus and ask what their father does, they'll say I work for NASA or I'm in the Air Force. You'd have to really dig to find out from them that I'm an astronaut."

Still, Lt. Colonel Guion Stewart Bluford, Jr., has a place forever in history as the first black American in space. He will be a hero for young people both now and in the future, and not just for those who want to become astronauts. He will be a source of encouragement for those who dare to dream big dreams and for those who are willing to work to make those dreams happen.

Asked what he has to say to kids, Guy Bluford answers, "They can do it. They can do whatever they want. There really aren't any barriers to hold them back. They may find it difficult at times—I had to struggle through those courses at Penn

State—but if you really want to do something and are willing to put in the hard work it takes, then someday—bingo, you've done it!"

Glossary

aerodynamics: the way the air moves around an aircraft's wings and body in flight

aerospace engineer: a person who designs craft such as space shuttles for traveling in space

artificial earth satellite: any man-made device put into orbit around the earth. The moon is a natural earth satellite. A space shuttle is an artificial earth satellite. Artificial earth satellites can be launched into orbit either by ground rockets or from a space shuttle's cargo bay.

asteroid belt: the group of thousands of small planets between Mars and Jupiter

atmospheric friction: the rubbing of the air against a spacecraft. This friction produces a fiery

Glossary

heat when a spacecraft first reenters the earth's atmosphere.

booster rocket: one of the two big engines that lift the space shuttle off of the launchpad

cargo bay: the big middle section of a space shuttle where large equipment, such as an artificial satellite, is carried

civil rights movement: Civil rights are the freedoms and rights that people have as U.S. citizens. Black Americans, who make up the largest minority group in the U.S., have been denied their full civil rights more than any other minority group. The civil rights movement was an effort made by many Americans to end discrimination against blacks and guarantee them their full civil rights.

cosmonaut: a Russian astronaut

debrief: to question an astronaut after he or she has returned to earth from space. The questioning is done by scientists trying to find out what the astronaut learned during spaceflight.

delta wings: the triangle-shaped wings on space shuttles and on some aircraft

deploy: to set an artificial satellite outside the cargo bay of an orbiting space shuttle

Glossary

electrode: a small electric device worn by astronauts on their heads which can tell scientists on the ground how the astronauts' eyes are moving

flight deck: the top area in the cabin of a space shuttle where the pilots sit and where some experiments are conducted

horizon: the edge of the earth that we see in the distance where the land meets the sky. In space the horizon is curved.

integrated: united, no longer separated. In the past, black children and white children were often not allowed to go to the same school. When those schools became integrated, both black and white children could attend them.

laser physics: an area of science which deals with a special kind of light and energy

mechanical engineer: a person who designs machines and tools

middeck: the middle area of the shuttle's cabin where many experiments are conducted and much equipment is stored

Mission Control: the people in a space center who direct the activities of astronauts in space

Glossary

mission specialist: a crew member in charge of experiments on a shuttle

orbit: the path followed by one body as it revolves around another. The earth orbits the sun. *Challenger* orbited the earth.

rotate: to spin like a top. The earth spins around, or rotates, one full turn every 24 hours. We call this rotation a day.

segregation: separation, the opposite of integration. When black children had to go to an all-black school and white children had to go to an all-white school, those schools were segregated.

shuttle simulator: a machine on the ground which is built to resemble the cabin of a shuttle and can be made to seem as though it is really taking off or flying

shuttle system: all of the equipment needed for a particular function of the shuttle, such as the system for heating and cooling the cabin space or the system for communicating with earth

solar system: the sun with the group of celestial bodies, such as Earth and Mars, that orbit around it

Glossary

space center: a place where astronauts train for spaceflight, from which spacecraft are launched, and where scientists keep track of spacecraft flying in space

Zero G: the condition of no gravity that causes the weightlessness of astronauts when they are in space

Index

aerospace engineering, 15, 17, 19, 22-23, 37
Air Force, U.S., 20-22, 38, 54
Air Force Flight Dynamics Laboratory, 22
Air Force Institute of Technology, 22-23
Aldrin, Edwin, 40
Apollo program, 40-41
Apollo 17, 8
Armstrong, Neil, 40
asteroid belt, 41
Astronaut Office, 39-40
Atlantic Ocean, 7

Banana River, 7, 16
black astronauts, 38-39, 43, 45, 54
black pilots, 12-13
Bluford, Lieutenant Colonel Guion S., Jr.
  accepted as astronaut, 37-40
  brothers of, 11, 13
  aboard *Challenger*, 7-9, 46-53
  childhood of, 11-18
  children of, 21-22, 38, 54
  at college, 19-21
  in high school, 17-18
  parents of, 11, 13-14, 18, 45
  race consciousness of, 13, 19
  receives pilot wings, 21
  in Vietnam, 22
Bluford, Guion S., III, 21-22, 38, 54
Bluford, James T., 21-22, 38, 54
Bluford, Linda Tull, 21, 38
booster(s), 9, 40
Brandenstein, Daniel C., 27

Cape Canaveral, Florida, 7, 16, 39, 45, 49
Cape Cod, Massachusetts, 21
cargo bay, 30, 50
Challenger, 7-9, 42, 44-45, 47-53
Chamberlain, Wilt, 45
Cheek, Dr. James, 45
civil rights movement, 19
Coleman, Bessie, 12
Columbia, 42
communications satellites, 42, 44, 48-49
Communist China, 21
Cosby, Bill, 45
cosmonaut, 20, 42

62

Index

delta wings, 23, 41
Dwight, Captain Edward, 38

557th Tactical Fighter Squadron, 22
Freedom 7, 20, 38

Gagarin, Yuri, 20
Gardner, Lieutenant Commander Dale, 27, 33, 35, 52

Hampton, Lionel, 45
Homestead Air Force Base, 24
Hope, Bob, 54
Houston, Texas, 37-38, 39, 40, 53
Huntsville, Alabama, 39

India, 44, 48-49
Indian Ocean, 52
Insat-1B satellite, 36, 49-50

Johnson, President Lyndon B., 38
Johnson & Johnson Pharmaceutical Company, 44
Johnson Space Flight Center, 24, 27, 28, 31, 37-38, 40
Jupiter, 41

Kennedy, President John F., 38
Kennedy Space Center, 39

Lawrence, Bob, 38
Little Rock, Arkansas, 13

McDonnell Douglas, 44
McNair, Ronald, 44
Mariner program, 41
Mars, 41
Marshall Space Flight Center, 39
middeck, 31

Mission Control, 8, 47, 50
mission specialists, 42, 44, 48
Mojave Desert, 52
moon, 40

National Aeronautics and Space Administration (NASA), 16, 37, 41-42, 53-54

orbit, 9, 16, 46, 50, 52
Otis Air Force Base, 21
Overbrook High School, 17

Pacific Ocean, 49-50, 52
Pennsylvania State University, 19-21, 54-55
Philadelphia, Pennsylvania, 11, 13
Pioneer program, 41

Reagan, President Ronald, 8, 51
Reserve Officers Training Corps (ROTC), 20-21
Ride, Dr. Sally K., 38, 42-43
Rockwell Aircraft Company, 39-40
Russia. *See* Soviet Union

satellites, 15, 42, 44, 48-49
Shepard, Commander Alan, 20, 38
Sheppard Air Force Base, 22
shuttle simulators, 8, 9, 40
shuttle systems, 7, 39
Skylab, 40
Snark missile, 16
solar system, 41
Soviet Union, 15-16, 20, 42-43
"space age," 16
"space race," 16, 20
space shuttles, 39-44. *See also Challenger*
Sputnik I, 16-17

63

Index

Thornton, William E., 27, 28, 35
Truly, Captain Richard, 27, 34, 35, 52

upper flight deck, 30

Venus, 41
Vietnam War, 20-22
Vostok I, 20

Williams Air Force Base, 21
women astronauts, 38, 42
World War II, 12
Wright-Patterson Air Force Base, 22

Young, John, 39

"Zero G" (zero gravity), 26, 46-49